# ONE SMALL SQUARE®

## *Coral Reef*

### by Donald M. Silver

### illustrated by Patricia J. Wynne

LEARNING
TRIANGLE
PRESS

*Connecting kids, parents, and teachers
through learning*

An imprint of McGraw-Hill

New York  San Francisco  Washington, D.C.  Auckland  Bogotá
Caracas  Lisbon  London  Madrid  Mexico City  Milan
Montreal  New Delhi  San Juan  Singapore
Sydney  Tokyo  Toronto

Every living thing pictured in this book can be found with its name on pages 40-43. If you come to a word you don't know or can't pronounce, look for it on pages 44-47. The small diagram on some pages shows the distance below the top of the square for that section of the book.

*For Sam Silver*

my unforgettable uncle.

Our appreciation to Dr. Alan Harvey of the American Museum of Natural History for his detailed comments on reef life; Marc Gave for sticking with this series; Ivy Sky Rutzsky and Maceo Mitchell for contributing their talents and time; and Thomas L. Cathey for his ongoing support.

Printed in the United States of America. Except as permitted under the United States Copyright Act of 1976, no part of this publication may be reproduced or distributed in any form or by any means, or stored in a database or retrieval system, without the prior written permission of the publisher.
Library of Congress Cataloging Number 97-074150
ISBN 0-07-057970-9

6 7 8 9 10 11 12 QDB/QDB 15 14 13 12

Picture fans, baskets, cups, stars, domes—endless shapes.

Think yellow, blue, purple, pink, gold—a rainbow of colors.

Whether you are at a coral reef or at home, always obey safety rules! Neither the publisher nor the author is liable for any damage that may be caused or any injury sustained as a result of doing any of the activities in this book.

Imagine stripes, dots, zigzags,
speckles, swirls—fantastic patterns.

Add in painted heads, false eyes, drummers,
sweepers, flashlights, clowns.

Put them all together and what
do you have? A circus? Maybe.
But a coral reef, for sure!

A coral reef is nature's underwater wonderland.
No circus can compare with it. No circus has more
dazzling costumes, more tricks, more death-defying
feats, or more scary moments. No circus is more
action-packed. Or more exciting.

3

There is a way, though, that a reef can't compete with a circus. A circus can come to your town; a coral reef cannot. Instead, nearly all reefs are very far away from where most people live. And all reefs are in the ocean.

But don't let that stop you from enjoying the magic of a reef. You can visit one right in this book.

So come along to the largest coral reef on earth—the Great Barrier Reef off the coast of Australia. Why "Great"? Because it stretches more than 1,250 miles (2,000 km), about the distance between Denver and San Francisco. Astronauts have seen this reef from outer space!

Why "Barrier"? Because it isn't found near land but out at sea. It sits there like a wall that can block a submarine from passing or sink a ship that hits it.

Nowhere else do so many kinds of flashy sea creatures live together. Nowhere else is it so hard to tell what is an animal, what is a plant, and what is a rock!

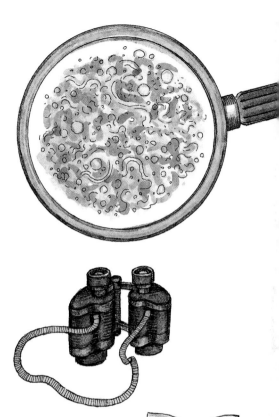

No coral reef nearby? Don't worry. With simple tools such as these, there are activities you can do to help you discover how nature works where you live. Reefs work in many of the same ways.

## What's Living on the Coral Reef?

Monera

Protists

Animals

# One Small Square of a Coral Reef

It's showtime on the Great Barrier Reef. Nippers, snappers, and ink shooters are on parade. Stingers, chompers, and grinders will soon take center stage. For the next 30 pages, one small square of coral reef will be the main attraction.

The show takes place in a space about as large as a four-person elevator. The square at the top is about 4 feet (1.2 meters) on each side. As you can see in the picture, the space below is about twice as deep.

Follow along as this book explores the wonders of the Great Barrier Reef. There will be activities you can do where you live to help you understand how reefs work all over the world.

So bring on the drummers, the puffers, and the damsels in distress. Keep an eye out for a giant and, of course, for hungry sharks. There will be plenty of snacks and even some clowning around.

Come one, come all! On with the matinee, the afternoon show. Is a coral reef truly the greatest show on earth? Dive right in, and soon you can decide.

Looking for the small square? It's in the box on the side of the reef that faces the open sea. On the other side, between the reef and land, is a lagoon.

# Top Billing

The water is warm and clear. Bright sun lights the sea. The waves are gentle. It's a perfect day to explore the small square—in diving gear.

Leave the air world behind. From here on, expect water—water everywhere. And get ready NOT to believe your eyes. One glance at the reef and you'll swear you see small trees, thick bushes, colorful flowers, and maybe even a cabbage patch. Don't be fooled. Like many magic tricks, these shapes are only illusions. The reef is no underwater garden.

Swim over and touch a "cabbage leaf": It feels as hard as stone. It should, for it is limestone rock. So is nearly the whole reef. Now, look closely at the stone. It's full of pits or valleys.

Reefs don't grow from seeds. They are built of countless stone cups made by hard corals. But soft corals also live in reefs. And there's not a stone cup on them. Each coral animal, hard or soft, is called a polyp.

Whatever you do, don't touch the "flowers." They're probably sea anemones, animals with stinging tentacles. In fact, there are millions of other stinging animals right in front of you. They are called hard corals. Most of them are hiding in the stone pits or valleys with their tentacles pulled back.

Most hard corals are no longer than your fingernail and no wider than poppy seed. They haven't a bone in their soft, saclike bodies Yet these small animals are the stars of the reef. They built it without ever lifting a rock. How? With the help of chemicals in seawater.

**Stinging cell**

**Alga**

**Tentacle**

**Limestone**

**Connection**

**Cup**

What can stinging tentacles, a mouth, two cell layers and some nerves do? How about form a hard coral that catches food and digests it, breathes, gets rid of wastes, and builds a stone cup? And has room for ultra-tiny food-and-oxygen-making algae to live inside it. In a second this hard coral will join the others it lives with and pull down inside its stone cup until it's ready to go to work.

9

## Your Reef Notebook

Start a notebook all about coral reefs. In it record the activities you do in this book. If you visit a pond, note what water creatures live there. Draw pictures of them. Can you figure out how they swim?

Read all about diving and take notes on what equipment and training you would need to reach a reef. If you come across pictures of reefs in magazines or newspapers, paste them in your notebook. Can you identify any of the animals? If you ever get the chance to explore a real reef, be sure to write down what it's like.

## What's in a Name?

No coral has a brain, not even brain coral. Why the name? Its shape. Each kind of coral grows into its own shape. What names would you give to the corals shown on these two pages? Did you match any of the names on page 41? Which do you like better? No doubt your names will win out for corals with only scientific ones.

## Block Party

Hard corals build a reef of stone a cup at a time. The shapes they grow into are amazing. All the animals inside each shape are connected. They are one colony.

Get out your Legos™ or other plastic building blocks. Let each block stand for a coral's cup. Build a "colony" of blocks in a shape you find in this book or in any shape you like. Invite your friends and have a party. Put all the "colonies" next to each other and make your own small square.

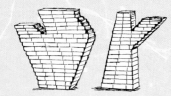

## Still Growing

How much do you grow in one year? How about a flowering plant in your backyard? Use a ruler and find out. Then start a growth chart in your notebook. If you have a measuring tape, wrap it around the trunk of a tree once a month. How much bigger does the trunk grow in a year? How do your findings compare with these? If a bush-shaped coral colony grows 6 inches (15 cm) a year, that's a lot. Many grow less than 0.5 inches (12 mm) per year.

If you think a three-ring circus is a lot to take in, a square of reef will make you dizzy. Animals are busy everywhere, trying to stay alive.

Your body made your skeleton by using minerals from the food you eat. A hard coral's body makes its skeleton by using chemicals it takes in from seawater. Your skeleton is inside your body. The coral's is outside. It's made not of bone but of stone that hardens from the chemicals. And it's cup-shaped, which means open on top. That way a hard coral can stretch out its tentacles to catch food. But it can't swim away—it's stuck to its cup.

Some hard corals live alone. But the ones that build reefs wind up with plenty of company. After one of these hard corals builds its cup, part of its body can squeeze off to one side. There it grows into an exact copy. The copy builds its own stone cup in such a way that the two stay connected. They share food and, faster than a fax, warn each other of danger. Also, they both can squeeze off more copies as new

Try to find the see-through shrimp. It's not easy for you or for predators.

With so many nooks and crannies, you'd think Christmas tree worms could find a place to live. They do—by making holes in the corals' limestone and moving in. If a worm is eaten, its home may soon be taken over by a fish or a crab.

neighbors. In this way, squeeze by squeeze and copy by copy, one tiny hard coral can become millions—all connected.

About a dozen kinds of hard corals live in and around the small square. One day, like all other animals, they die, perhaps by being eaten. When their bodies are gone, their skeletons will remain. Other hard corals can build their cups on top of them. That's right, the small square of today is built on top of the skeletons left behind by hard corals that lived hundreds, even thousands, of years ago. Top that!

There's oxygen in the water, but you can't breathe it. Reef life can. Fishes have gills to take in oxygen. Their blood then carries the oxygen to all body parts. Each cell of a coral animal's body soaks up oxygen directly from the water.

# Damsel in Distress

A little blue damselfish keeps poking in and out of a hole. Is it in distress? Mad as a hornet is more like it! Suddenly, it darts out and attacks a surgeonfish almost ten times its size. The big fish quickly gets the message. In a flash it is gone. For the fearless damsel such a daring feat is all in a day's work. It must do what it takes to protect its turf.

The turf is a patch of seaweed—a kind of algae. Like plants, algae capture energy from the sun and use it to make food and oxygen. The turf is the damselfish's breakfast, lunch, and dinner. Small shrimps and crabs can nibble too, but any big algae eater that tries to muscle in has to face the damsel's fury.

Don't worry, there's plenty more algae in the small square for other fishes, sea urchins, and worms to eat. Some float in the water. And lots grow around and between the hard corals' stone cups. But the algae that live inside the hard corals' bodies are off limits. They are protected by the stone cups and the stinging tentacles of their hosts.

That's good for the hard corals. Without eating their algae, the corals get to share some of the food and oxygen. These help the corals grow and keep building their cups. In return, the algae recycle the hard corals' wastes. With these, the algae can stay healthy and make more food.

Just when you thought no coral could move, a fungus coral drifts off. Just when you thought all corals close by day, one opens (see circle).

Tail first, a pearlfish wiggles into its home—the rear end of a sea cucumber.

12

Some shrimps, snails, barnacles, and crabs never leave this soft coral's bendable branches. Can you find any of these creatures? They are so small that you may need a magnifying glass.

A damselfish's seaweed patch may serve only as food. But the algae growing around coral cups also give off limestone. It strengthens the coral reef by cementing cups together and filling gaps.

# The Reef Giant

Who isn't awed by a giant? There are no human giants like the ones in movies who have one eye and swing great clubs. Nor are there giant apes who climb the Empire State Building. But there are giant trees taller than a 25-story office tower. There are whales with tongues the size of a minivan. And there is the giant of the reef, the giant clam.

Like other clams, the giant lives inside its hard shell. Some giant shells grow more than 3 feet (1 m) across and 2 feet (0.6 m) high, and weigh over 500 pounds (230 kg). That's big enough for you to sit in. Compared with clam shells on a beach, it's huge.

The giant never leaves its wavy shell. When the shell opens, two tubes stick out of colorful, fleshly "lips." Guess what's living in the lips? Algae. Like hard corals, the giant

Sometimes it pays to have a big mouth, sometimes it doesn't. It all depends on what you eat. Look at every mouth in the circles on these pages. Can you figure out how each is suited to doing its job of taking in food?

What a sloppy eater the parrotfish is! It's after algae, yet it bites off hard corals too, stone cup and all. The fish eats the coral and grinds the cup into sand.

14

Eyespot

Mantle

While you are watching a giant clam, the eyespots in its mantle see your shadow.

clam soaks up part of the food the algae make. The clam also eats extra algae as they multiply. But that's still not enough to fill a giant. Here's where the tubes come in. One pulls water into the clam's body so it can pass over its gills. The gills remove oxygen to breathe and tiny, floating food to eat. As the gills pass the food to the giant's hungry mouth, the water flows out the other tube.

With such a simple diet, few reef animals need fear the giant clam. Divers certainly are in no danger. After all, clams don't eat people—people eat clams. But a little awe never hurts to keep a giant clam in its shell and off someone's dinner plate.

Don't get suckered into thinking the little blue-ringed octopus is cute. Sure, it has suckers on its arms for crawling slowly about the square and grabbing crabs or fishes to eat. But its bite is full of poison that is deadly to its prey— and to people.

15

The boxfish and the scorpion fish are dressed to kill. They are poisonous; so deadly that they don't have to be good swimmers. They may even stay around a diver. Although you could catch one, you NEVER want to.

Wouldn't you know it? Someone always wants to get in on the act. This time it's a cleaner shrimp targeting pests on a coral trout's back while a cleaner fish zeroes in on dinner in the trout's open mouth.

# Swim Right Up

Tightrope walkers use a net. So do daring trapeze artists, and fliers shot from a cannon. When work is that risky, safety comes first. Yet in the small square, cleaner fishes throw caution to the waves and swim into the jaws of death without a safety net in sight.

It starts with the way a little cleaner fish moves. Its dip-and-wiggle dance signals bigger fishes to swim right up so the cleaner can go to work. One by one the cleaner's clients get into line. Without pushing or fighting, each big fish peacefully awaits its turn, as if under a spell.

The cleaner wastes no time. It nips at the skin of the first in line. It picks at lifted gill covers and at fanned-out fins. And when the big fish opens its mouth, the cleaner swims in to nip some more. What a trap! Surely it's curtains for the cleaner.

No—the cleaner is never in danger. By letting the cleaner do its work, the big fish gets rid of tiny bloodsucking and skin-eating pests that might make it sick. And the pests just happen to be a cleaner's favorite foods. For its part, the cleaner gets its meals delivered directly to its cleaning station. That's the spot where the cleaner sets up shop and cleans whatever swims right up—divers, too!

Wash your hands. Take a bath. Time for a shower. No doubt you hear such commands every single day. The reason is simple: Keeping clean keeps you healthy. It gets rid of many bacteria and germs that could make you itch, turn cuts into infections, or sicken you.

Germs and pests attack animals too. Fishes on a coral reef are lucky there are cleaning stations. Zebras in Africa let tick birds land on them to get rid of nasty ticks.

Watch your cat, dog, or other pets. How do they groom themselves? What about birds and squirrels? Use binoculars to get a close look at them. Write down what you see and draw pictures in your notebook. If you visit a zoo and happen to catch other kinds of animals grooming, add them to the list.

17

# Costume Change

Have you ever worn a costume in a play? Or on Halloween? Perhaps the disguise was so perfect that no one knew who you really were. Maybe it helped you hide and scare your friends. A costume can make you look silly, serious, or anything in between. It's all in the shape, colors, and patterns of what you wear. The same is true for many reef animals. If you think their looks never change, think again.

Take the cuttlefish and its cousin, the octopus. These quick-change artists flash new body colors when they are angry. That's often enough to send other animals fleeing.

He's got all the right moves. He's flashing all the right colors too. In cuttlefish talk, he's signaling her: "Let's mate."

This false cleaner fish looks just like the real one. Instead of cleaning, though, it bites out chunks of flesh and flees.

This costume is a winner if you can't tell where the fish ends and the coral begins.

Then there's the young spotted sweetlips. When this fish swims, its pattern stands out. But when it rests on a patch of coral, it seems to disappear because it blends in so well with the stone. And what about the reef crab, which lifts a sponge onto its shell so crab eaters will mistake it for the bad-tasting sponge and leave it alone.

How a predator looks can help it sneak up on its prey unseen. But if, in an instant, the prey can change the way it looks, it may be able to confuse a predator just long enough to make an escape. Check out the "costumes" worn in the small square. If you were a predator, which would most help you ambush a tasty meal? If you were prey, which would most help you if your life was in danger?

Will the real eel take a bow? Will the fish that looks like a sponge show itself? Will another fish reveal it's not really poisonous? Not if they all want to stay alive.

Sponge off a sponge? Sure. There's room to spare for animals to hide down inside.

# Clowning Around

If fishes could talk, here's what clown-fishes might say: "Don't we look tasty? Come and get us. We dare you. We double-dare you!" But there would be few takers. Any fish eater that took the dare would be in for a nasty surprise.

A clownfish spends most of its life in the center ring—of a sea anemone's waving tentacles. The wily clown briefly steals off for a snack of shrimps. But it always quickly returns to the center ring.

Practice makes perfect—only this isn't practice. It's slime time. Young clown-fishes must swim back and forth and rub against an anemone's tentacles. That way slimy mucus from the anemone can stick to their own. Without slime they would be stung.

If something touches the trigger, FIRE— a stinging cell springs open and out shoots a spearlike thread.

Trigger

Stinging cell

Should a predator be foolish enough to follow the clownfish home, ZAP. Every tentacle the predator touches shoots out spiny, stinging threads. Any that stick into the predator's skin can inject poison that stuns, numbs, and may even kill the predator. When the tentacles reel in their catch, they pass it along to the anemone's mouth. Uneaten morsels don't go to waste: The clownfish gets its chance to feast.

But wait: Something fishy is going on. The anemone is a fish eater yet never eats its tasty clownfish. That's because slimy mucus is part of the clown's bag of tricks. The mucus coats the clown's body and shields it from the deadly stings. Without the slime, there would be no clowning around.

Blue coral

Black coral

Mouth

Tentacle

Living blue coral is brown, and living black coral is yellow or orange. It's only the skeletons underneath the animals that truly live up to the colors of their names.

Mouth, tentacles . . . no it's not another coral animal. It's a sea anemone that clings tightly to stones or shells. Anemones come in many colors, but all have stinging cells to catch food or defend against predators.

A cave can be a fish's trapdoor for a fast exit from danger.

# Sideshow

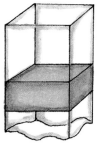

Part of the reef is missing. Not far below the square, it is definitely gone. Maybe the hard corals just grew that way. Maybe storm waves crashing down pulled out a big block of coral rock. Or maybe some creatures ate away all the limestone cups. No matter: The hole has taken on a life of its own as a coral cave.

From the outside looking in, the cave is dark and spooky. If you shine a light, you can see animals almost everywhere. Stuck to the cave's roof are hard corals that don't love sun as reef builders do. They are loners without even algae inside to keep them company. Nearby hang seaweeds that don't make their own food. And that's just for starters.

This sideshow boasts sponges and sea squirts, a feather star with a maze of branching arms, and a school of bright-red cardinalfishes. There are soft corals, too, loaded not with stony cups but with tiny, colorful limestone needles.

What lives in the cave? What popped in to hide? Or to snuggle down for a day's snooze? It's really hard to tell. But all that is about to change.

## Stone breakers

Boring urchins

Boring sponges

Mollusks and worms

Hard corals are stone makers, but the square is also home to stone breakers. Some sponges and worms scoop out holes in the limestone and move right in. Sea urchins rely on their chisellike teeth to crunch away rock. What a mouthful!

A feather star's delicate arms
easily break off. But the star
has a trick up its sleeve. It can
grow new ones.

The coral trout slips into the
cave to nap the day away.
Yes, it's asleep. If it had
eyelids, it would close them.

Mouth

Tentacles

Stomach

Sea squirt

You can call it squirt,
but watch out all the
same. A sea squirt can
squeeze its muscles and
squirt a jet of water out
its mouth.

Soft coral
polyp

23

Presto! Right before your eyes a butterflyfish changes its colors and patterns to better hide among the corals.

Here come the plankton—very tiny animals and simpler creatures that rise at night from nooks and crannies.

Plankton

# Disappearing Act

For a few seconds the small square all but vanishes. It is overrun by a school of fishes and looks more like rush hour on a freeway than part of a coral reef. Suddenly, as if out of nowhere, a reef shark attacks. There is chaos as the fishes spread out in alarm. But not all are quick enough to escape the gaping jaws of the hungry predator.

The shark arrived with perfect timing. The fishes, too, were right on cue. When the sinking sun signals day will soon be done, you can be sure the disappearing act has already begun.

The school is on its way out to sea to spend the night. Other fishes grab a last bite of algae before

Another opening, another show. One by one, hard and soft coral polyps uncurl their tentacles.

What's that smell? Not a parrotfish, once it wraps up inside a see-through bubble of mucus. The bubble traps scents so no hungry eel or other predator can pick them up.

nestling in among coral branches or down inside sponges. Their colors change to help them blend in and confuse most night hunters. Anemones pull in their tentacles and close up shop, with their clownfishes safely tucked in. Every creature seeking shelter is in a hurry to find a hiding place while it is still light. And all must beware of reef sharks, groupers, and other predators that see well as darkness comes on.

Meanwhile, back in the cave, the feather star is making its move. The cardinalfishes are ready to swim out into the square. The matinee is at an end. The night performers can no longer wait in the wings. Their moment has come.

Muscular foot

# Lights!

Soon after nightfall the small square is ready for its evening show. Stars curl and uncurl their arms. The fans are out. A Spanish dancer glides along like a surfer riding a wave. Drummers, sweepers, crawlers, and creepers are all in motion. The lion and the tiger hunger for more food. And everywhere you look, spines, stalks, branches, and tentacles are swaying.

There's just one hitch. The small square is in almost total darkness. If you want to see the show, you must supply the light. A waterproof torchlight could cover the entire square, as it does in this book. But a flashlight is easier to handle, and its beam is a perfect spotlight on the creatures of the night.

It takes hours and hours of practice before a diver is ready to explore a coral reef at night. Even then, no one should ever dive alone but only with an experienced buddy who knows the reef, how the water moves, and how to be extra careful not to get hurt.

The reef is a wild place, so prepare for a scare or two. Or even a fight to the finish. It's all in a night's work for animals trying to make a living. But whatever you do, be sure not to miss the grandest finales of all. Lights!

Look at those sea slugs. They're not showing off. These snails without shells creep, crawl, ripple, and glide around the small square, nibbling sponges and sea anemones. Many of the slugs get their colors from what they eat.

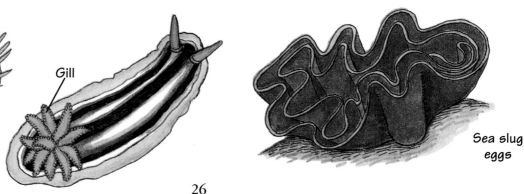

Gill

Sea slug eggs

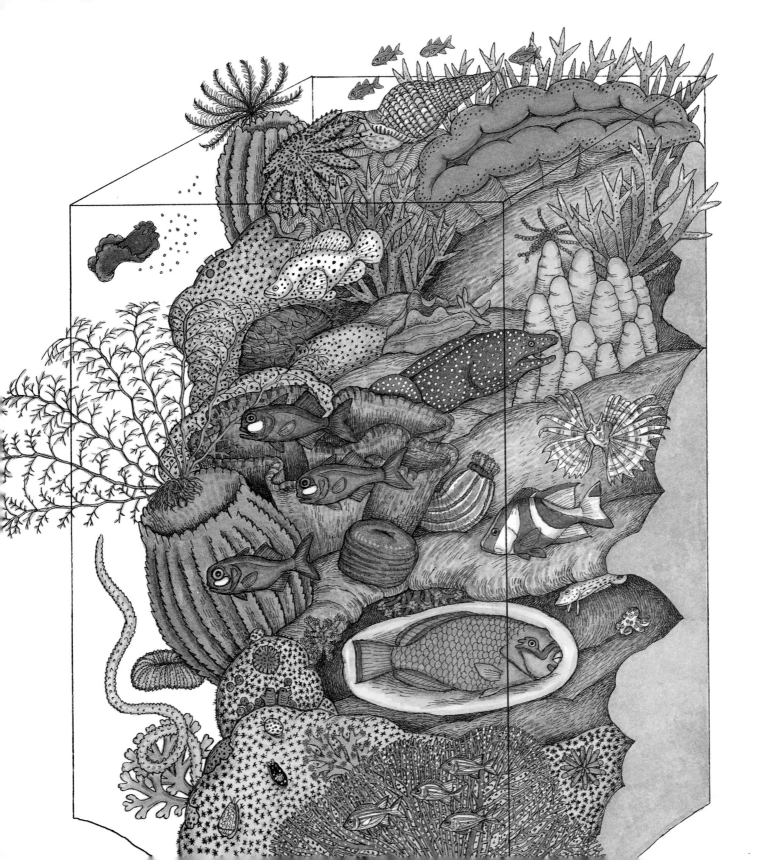

# Coral Wars

Imagine how changed a garden would look if every flower bloomed at the same moment. Or a skyscraper if flags were unfurled from every window. That's how changed the small square is at night. The limestone goes fuzzy with feeding corals and uncountable numbers of tentacles all stretched out. Waving and stinging, the tentacles catch plankton. They fill the corals' mouths with shrimplike krill, baby fishes, tiny eggs, and other seafood floating by in the moving water.

And then there's the sweeper, a long, deadly tentacle out not for food but blood. If it reaches over and touches

Nothing—not even their stony cups—can save corals from the crown-of-thorns' stomach. Not when it comes out of this starfish's mouth and makes a meal of corals.

A search-and-destroy mission is a double success when deadly stomach threads not only kill other corals but eat them.

In war, some win, some lose. This brain coral is losing as encrusting corals grow on top of it.

Spanish dancer

28

One coral colony may lose the sunlight if another grows tall and blocks it.

Bacteria, too tiny to be seen, live in reef animals and on stones.

an enemy—a coral animal not part of its own colony—the enemy must be killed. That's the only way to stop another colony from slowly invading and smothering living polyps. Or to keep a different kind of coral from growing too close for comfort and blocking the flow of floating food.

When one colony declares war, it may soon find itself under attack. The enemy may fight back with sweepers, poisonous mucus, or stomach threads that can seek out and destroy.

Don't expect a truce anytime soon. Eating can keep a coral colony alive from day to day. It's the war over space that a colony must win to hold on to a place in the square.

No-man's land

Sweeper tentacles

Sweeper tentacles fly as one hard coral attacks and another fights back. The battle zone becomes a no-man's land where nothing lives.

On
Off

Glowing
bacteria

Like fireflies, flashlight-
fishes shine at night.
Their eye pouches are
full of bacteria that
glow. When special
muscles fold down the
pouches, it's lights out.

Cone snail: Ready, aim, fire!
Harpoon that tiger cowrie
with a poison dart. Then
pull it in and eat it.

Dart

# In the Lion's Mouth

Reef animals—and divers too—make way for the lionfish, if they know what's good for them.

Watch the lion spread its fins wide. See the lion sweep small fishes along if they don't hurry aside. Marvel as the lion herds the fishes into a corner, opens its mouth, and sucks them in like a vacuum cleaner. Wait, if you want, for small fishes to swim back out—only they won't. They aren't protected like cleaner fishes. They are the lion's dinner and you can't save them. If you try, the lion might stab you with one of its dagger-sharp spines. Stay away from those spines at all costs. They can be very painful.

Night on the reef belongs to predators. These hunters must kill to stay alive. Some can make out prey in the dimmest glimmer of light. Others smell or taste the water for food. Still others feel around for their next meal. There are dart shooters and netters and powerhouses that use sheer muscle-might to pull open shells.

Just about every kind of creature in the small square is hunted. As long as there are vast amounts of plankton, there will be more than enough small fishes and other plankton eaters to fill big mouths, including the lion's.

## Talent Scout

You may find more than a hundred kinds of fishes in one small square of coral reef. That's because more kinds of fishes live on reefs than in any other part of the ocean. Add in corals and other reef creatures and what do you have? An incredible variety of talents for the matinee and evening shows.

How does your backyard talent compare? List in your notebook every living thing you scout out there. Use a magnifying glass to view tiny creatures on leaves, flowers, and in the dirt. To find out what each one is, use a field guide.

A field guide contains the names of plants and animals that live in different places. There are field guides to fishes, seashells, birds—just about everything in nature. If you don't have field guides at home, try the library. Start your own coral reef field guide. Draw pictures of reef life or paste in photos you find. Rent videos about the Great Barrier Reef or other reefs, and use your guide to try to identify what you see.

## Moon Watch

When reef waters swirl with sperm and eggs, you can be sure there was a full moon a few days before. The moon, the sun, the seasons, and the weather affect what many living things do. Do birds leave your town in autumn and return in spring? Try to figure out when. After the first few cool or warm nights? After a full moon? Before or after leaves begin to fall or to grow?

Each year when do you see the first bees? Different flowers? Caterpillars? You may uncover a connection no one else ever noticed.

## Front-Row Seat

It may be a while before you have the chance to explore a coral reef. Or before you learn how to dive carefully so you harm no reef creatures. Until then, you can still get a front-row look at a living reef in an aquarium. If there is one near you or near a place you vacation, be sure to stop by. Bring along your notebook and draw pictures. Jot down animals' names if they are listed. Ask aquarium guides who cares for the reef and how.

Like an erupting volcano, a milky cloud of sperm spews out from a barrel sponge. Nearby sponges draw in the sperm to reach their eggs.

For some fishes the "grand finale" is the right time to squirt out sperm and eggs as they mate.

## The Grand Finale

Once a year the reef puts on a show unlike any other. The timing is out of this world—a few days after a full moon late in spring. As the sun sets, all is ready. Inside coral polyps, colorful round bundles are in place.

Just after dark, the first polyps open their mouths. Out float the bundles. There's no turning back now. Within seconds, all the other polyps within that colony let go with theirs. By then, other kinds of corals start to join in. Some let their bundles loose a patch at a time instead of all at once. Soon the water is clouded by a blizzard of bundles. But the show's not over. It's just beginning as the bundles burst open like fireworks.

Sperm, eggs, or both stream out of the broken bundles. The eggs float while the sperm swim about like tadpoles. There are so many eggs and sperm mixed together that the hungry predators feasting on them still leave vast numbers behind when they have had their fill.

That's the way it must be if the small square and the rest of the reef is to have a future. The job of each kind of sperm is to find an egg of its own kind and unite with it. Then the egg can grow and hatch into a baby coral animal—if it is not eaten.

Catch this act: To confuse a predator, an octopus shoots ink, then speeds away.

Bundle
1

Sperm
Egg
2

3

4

5

From the corals come colorful bundles (1) of sperm and eggs (2). From sperm and eggs come new corals (3) that swim and attach to shells or stones (4). After building its cup (5), a hard coral can grow into a colony that gives off its own colorful bundles.

This male mushroom coral gives off only sperm. The females give off the eggs.

A baby coral will swim in the plankton. If it's not eaten, it may settle down one day on top of a dead coral, on an empty shell, or on some other hard place. If that hard place is in clear, warm waters, the coral may be able to build a stone cup—if it's not eaten. If the coral polyp in its cup is not attacked by other hard corals, it may grow into a colony of its own, squeeze by squeeze, copy by copy, stone cup by stone cup.

From the billions upon billions of sperm and eggs that take part in the grand finale, the small square may wind up with one new colony of hard corals. Who knows? Sometime in the future, that colony may be one of the Great Barrier Reef's star attractions.

At dawn, that colony will pull in its tentacles just as the hard corals do in the small square today. Sharks and other predators will be on hand to prey on night life trying to safely disappear. Or on any day animals that come out of hiding before they are fully awake and ready to repeat their acts. On with the show!

What makes breaking waves sparkle and glow at night? The answer: Many sea creatures that light up. Are they signaling each other? Trying to attract prey? Warning of danger? Scaring off predators? No one yet knows.

Spanish dancer and green spotted sea slug

Plankton

Golden coral

Encrusting sponges

Encrusting bryozoan

Dendro-nephthya

---

## One Small Square of Coral Reef at Night

Can you match each living thing with its outline?

Fire coral

Black coral

Textile cone snail

Tiger cowrie snail

Sea squirts

Fungus coral

Rosy pink cowrie snail

Squirrelfish and sea fan

Tube worms

Crown-of-thorns
and triton snail

Acropora
branching coral
and damselfishes

Feather star
and barrel sponge

Brain coral

Barramundi
cod

Giant
clam

Columnar porites
coral

Leafy
montipora
coral

Whitemouthed
moray eel

Halimeda coralline
algae

Slate pencil
urchin

Speckled
sand perch

Flashlight fish

Organ pipe coral

Massive porites
coral

Lionfish

Blue-ringed
octopus

Parrotfish in
mucus bubble

Sea
anemone

Basket star

Violet
sea
cucumber

Red emperor

## Reef in a Box

Measure the length and the height of a shoe box. Cut a piece of paper for the background wall about ¼ inch (6 mm) shorter than the box height and about 4 inches (10 cm) longer than its length. Draw hard and soft corals on it, along with sponges and other reef life. Place the picture in the box and tape each side to the front. The picture will curve.

With Legos™ or blocks, build coral shapes in front of the picture. On separate sheets of paper draw reef fishes and other animals, each with a flap. Cut out each picture, bend its flap, and glue or tape it to the box or to the corals you built. Cut out some fishes without a flap and hang them with taped string from the top of the box. Create a daytime reef and a nighttime one.

## Can you match each living thing to its outline?

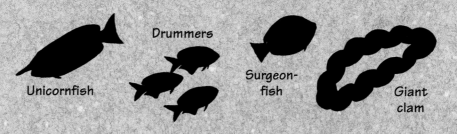

Unicornfish   Drummers   Surgeon-fish   Giant clam

Acropora branching coral and damselfish

Brain coral

Sea anemone and clownfish

Butterflyfish

Blue-ringed octopus

Sea cucumber and pearlfish

# Give Reefs a Chance

There's still so much to discover in one small square of coral reef. And the rest of the Great Barrier Reef awaits with new and exciting wonders playing night and day.

But not everyone who visits a reef is there to explore without doing harm. Many people are destroying parts of reefs by overfishing, collecting live animals, gathering coral stones, drilling for oil, mining, diving carelessly, and not watching where they sail ships. Reefs the world over are also in danger from pollution that can dirty or overheat the water and make coral animals sick. Then some corals toss out their algae and turn white as ghosts. They soon become so weak that whole colonies die.

More and more countries are turning their reefs into national marine parks protected by laws. That's the only way to give reefs a chance to recover from damage already done and to save them in the future. You can help by not buying live animals taken from reefs or products made from anything that once lived on a reef. Let's keep earth's greatest show from ever closing.

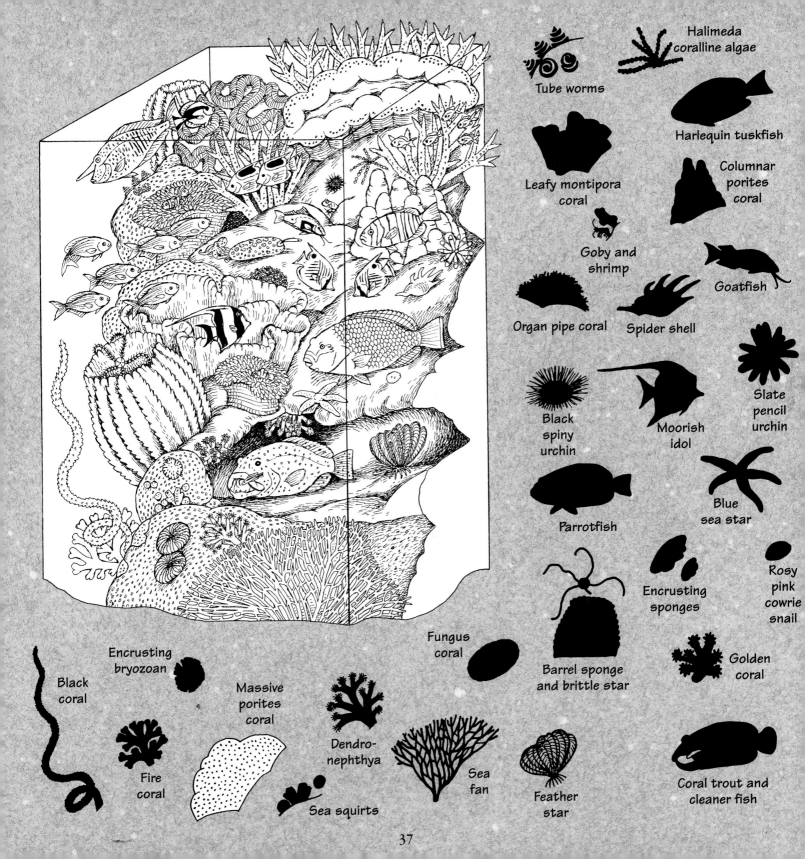

Tube worms

Halimeda coralline algae

Harlequin tuskfish

Leafy montipora coral

Columnar porites coral

Goby and shrimp

Organ pipe coral

Spider shell

Goatfish

Black spiny urchin

Moorish idol

Slate pencil urchin

Parrotfish

Blue sea star

Encrusting sponges

Rosy pink cowrie snail

Barrel sponge and brittle star

Golden coral

Fungus coral

Black coral

Encrusting bryozoan

Massive porites coral

Dendro-nephthya

Sea fan

Feather star

Coral trout and cleaner fish

Fire coral

Sea squirts

# Reef Finder

Looking for a reef? Try the parts of the sea shown on the map in blue-green. There the water stays warm all year—around 68–77° F (20–25° C). That's also where the water is very clear and sunlight easily reaches the algae that live inside hard corals.

You'll find most reefs within 50 feet (15 m) of the surface. If you dive, be ready for waves and currents that keep water flowing over the reef all the time.

Atlantic Ocean

Hawaii

Florida

Caribbean Sea

Sea of Cortez

Pacific Ocean

Hawaii's reefs may be small, but they are home to some of the world's most stunning fishes.

In the Sea of Cortez a green sea turtle may swim by a coral reef. So may a mighty humpback whale.

Millions of people live within a day's drive of Florida's coral treasures.

Together, the Indian and Pacific oceans are vast. Their reefs—including the Great Barrier and Hawaiian—are spectacular.

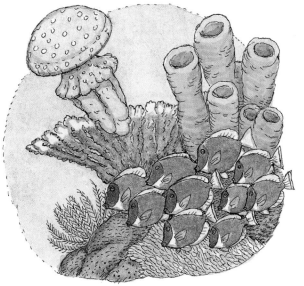

Pacific Ocean

Red Sea

Indian Ocean

Great Barrier Reef

You'll be amazed at how many reef animals live in the Red Sea and nowhere else. And what a show they put on every hour of the day.

Sunfish

Some Caribbean reefs are among the world's most colorful. Their corals grow tall and their fishes big. At 10 feet (3 m), the sunfish is a visiting giant.

Fishes are vertebrates—animals with backbones. So are reptiles, such as sea turtles; mammals, such as whales; birds, and amphibians. Look for seabirds hunting in reef waters. Expect no amphibians, because they can't live in the salty sea.

## Fishes and Other Vertebrates

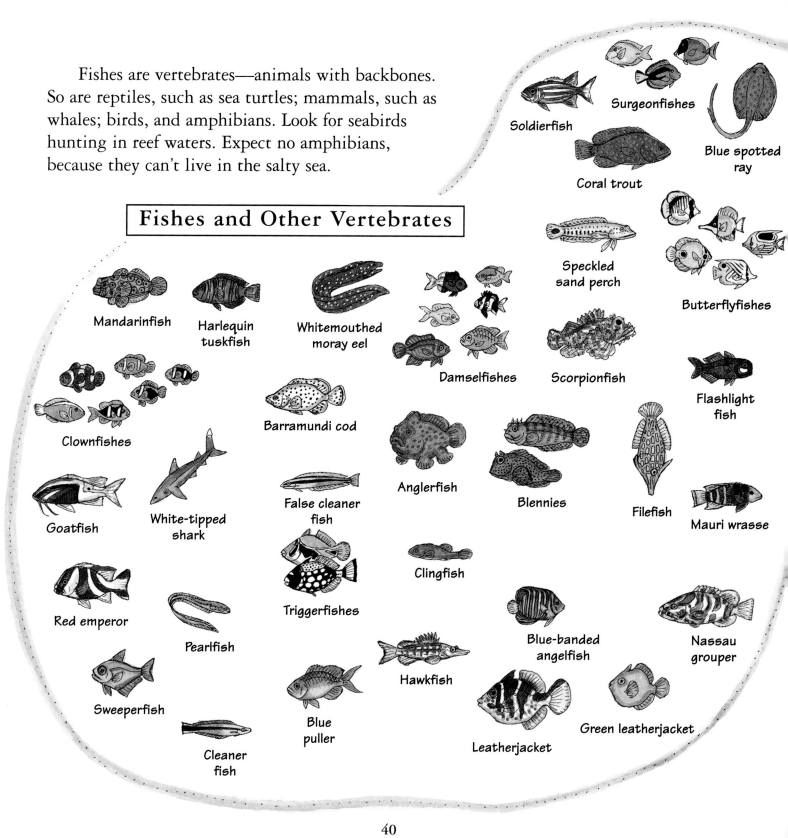

Soldierfish

Surgeonfishes

Blue spotted ray

Coral trout

Speckled sand perch

Butterflyfishes

Mandarinfish

Harlequin tuskfish

Whitemouthed moray eel

Damselfishes

Scorpionfish

Flashlight fish

Clownfishes

Barramundi cod

Anglerfish

Blennies

Filefish

Mauri wrasse

Goatfish

White-tipped shark

False cleaner fish

Clingfish

Red emperor

Triggerfishes

Blue-banded angelfish

Nassau grouper

Pearlfish

Hawkfish

Sweeperfish

Blue puller

Green leatherjacket

Leatherjacket

Cleaner fish

40

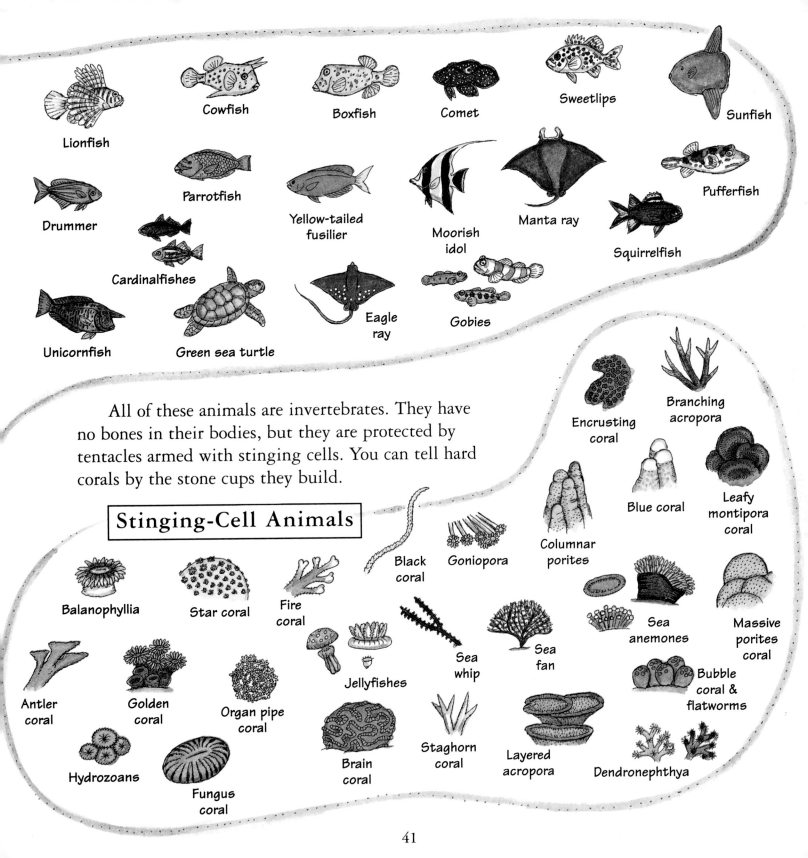

Lionfish

Cowfish

Boxfish

Comet

Sweetlips

Sunfish

Drummer

Parrotfish

Yellow-tailed fusilier

Moorish idol

Manta ray

Pufferfish

Cardinalfishes

Squirrelfish

Unicornfish

Green sea turtle

Eagle ray

Gobies

All of these animals are invertebrates. They have no bones in their bodies, but they are protected by tentacles armed with stinging cells. You can tell hard corals by the stone cups they build.

## Stinging-Cell Animals

Balanophyllia

Star coral

Fire coral

Black coral

Goniopora

Encrusting coral

Branching acropora

Blue coral

Leafy montipora coral

Columnar porites

Antler coral

Golden coral

Organ pipe coral

Jellyfishes

Sea whip

Sea fan

Sea anemones

Massive porites coral

Hydrozoans

Brain coral

Staghorn coral

Layered acropora

Bubble coral & flatworms

Dendronephthya

Fungus coral

Most of the animals living on reefs have no bones. Like corals, they are invertebrates. Look for crabs with jointed legs, snails with and without shells, and spiny-skinned animals such as sea stars, sea urchins, sea cucumbers, basket stars, and brittle stars.

## Other Invertebrates

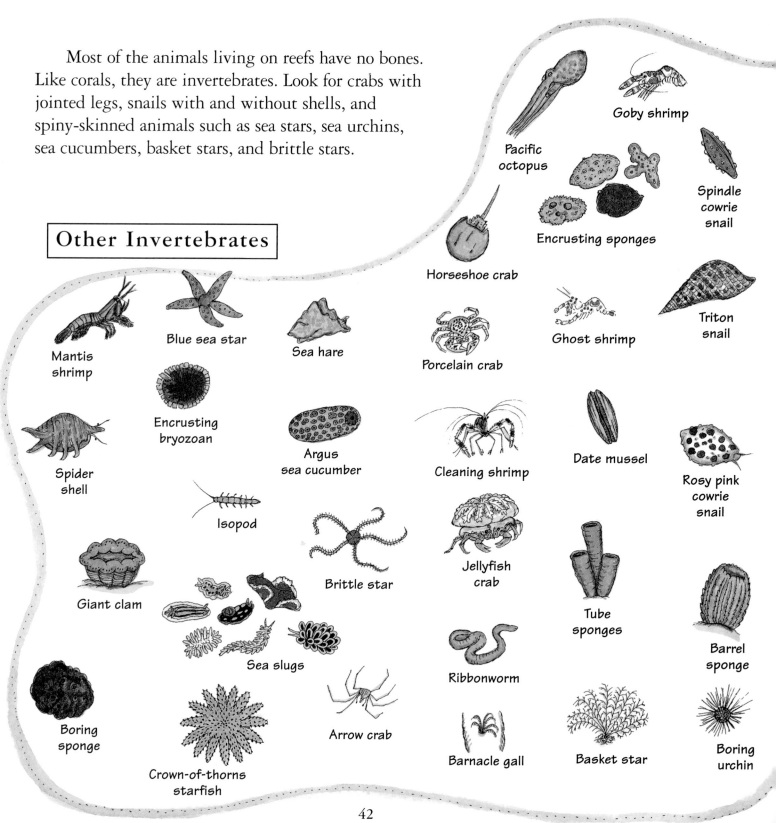

Goby shrimp

Pacific octopus

Spindle cowrie snail

Encrusting sponges

Horseshoe crab

Ghost shrimp

Triton snail

Mantis shrimp

Blue sea star

Sea hare

Porcelain crab

Spider shell

Encrusting bryozoan

Argus sea cucumber

Cleaning shrimp

Date mussel

Rosy pink cowrie snail

Isopod

Brittle star

Jellyfish crab

Giant clam

Sea slugs

Tube sponges

Barrel sponge

Boring sponge

Crown-of-thorns starfish

Arrow crab

Ribbonworm

Barnacle gall

Basket star

Boring urchin

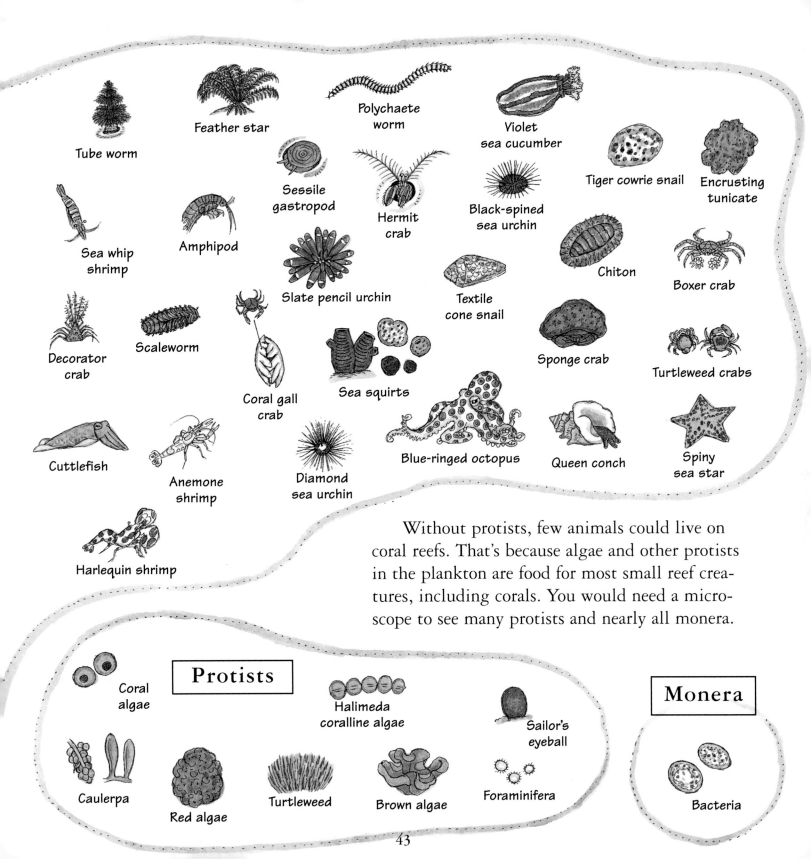

Tube worm

Feather star

Polychaete worm

Violet sea cucumber

Tiger cowrie snail

Encrusting tunicate

Sessile gastropod

Hermit crab

Black-spined sea urchin

Sea whip shrimp

Amphipod

Slate pencil urchin

Chiton

Boxer crab

Textile cone snail

Decorator crab

Scaleworm

Coral gall crab

Sea squirts

Sponge crab

Turtleweed crabs

Cuttlefish

Anemone shrimp

Diamond sea urchin

Blue-ringed octopus

Queen conch

Spiny sea star

Harlequin shrimp

Without protists, few animals could live on coral reefs. That's because algae and other protists in the plankton are food for most small reef creatures, including corals. You would need a microscope to see many protists and nearly all monera.

**Protists**

Coral algae

Halimeda coralline algae

Sailor's eyeball

Caulerpa

Red algae

Turtleweed

Brown algae

Foraminifera

**Monera**

Bacteria

# Index

## A

**algae** (AL-jee); *singular* **alga** (AL-guh) 9, 12, 13, 14, 15, 22, 25, 36, 38, 43. *Kinds of protists that can make their own food.*

**amphibian** (am-FIB-ee-in) 40. *Bony animal that lives the first part of its life in water and the second part on land.*

**animal** 5, 26, 29, 31

**aquarium** 32

**arm** 15, 22, 23

**Australia** 5

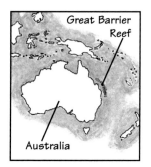

## B

**backbone** 40

**bacteria** 17, 29, 30. *Kinds of monera—one-celled creatures that don't have a nucleus (control center).*

**barnacle** 12

**basket star** 42

**binoculars** 17

**bird** 40

**black coral** 21

**blood** 11

**bloodsucker** 17

**blue coral** 21

**bone** 9, 10, 41, 42

**boxfish** 16

**brain coral** 9, 28

**breathing** 9, 11

**brittle star** 42

**butterflyfish** 24

## C

**cardinalfish** 22, 35

**Caribbean Sea** 39

**cave** 22, 25

**cell** 9, 11. *Smallest living part of all plants, animals, and funguses. Some living things are made up of just one cell.*

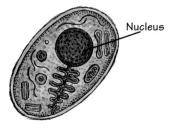

Nucleus

**Christmas tree worm** 11

**clam** 14

Inside a clam

**cleaner fish** 16, 17, 31

**cleaner shrimp** 16

**cleaning station** 17

**clownfish** 20, 21, 25

**colony** 10, 29, 32, 33, 34, 36

**cone snail** 30

**coral animal** 8–14, 19, 22, 24, 28 29, 32, 33, 34, 36, 39, 41, 42, 43

**coral trout** 16, 23

**crab** 11, 12, 13, 15, 19, 42

**crown-of-thorns starfish** 28

**cup** 8, 9, 10, 12, 14, 22, 28, 34, 41

**cuttlefish** 18

## D

**damselfish** 12, 13

**diving** 8, 9, 16, 17, 26, 32, 36, 38

## E

**eel** 19, 25

**egg** 28, 32, 33, 34. *To scientists, an egg is a female reproductive cell in a plant or an animal.*

# Index

encrusting coral 28
energy 12. *Ability to do work or to cause change in matter.*
eye 15

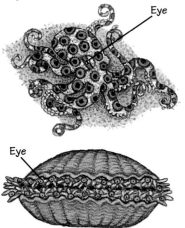

Eye

Eye

eyelid 23
eye pouch 30

**F**
false cleaner fish 18
feather star 22, 23, 25

field guide 31
fin 17, 31
fish 11, 12, 15, 17, 19, 20, 22, 24, 25, 28, 30, 31, 36, 38, 39, 40
flashlight 26
flashlight fish 30
Florida 38
food 9, 10, 12, 15, 21, 22, 26, 29, 43
fungus coral 12

**G**
germ 17
giant clam 14, 15
gill cover 17

Fin

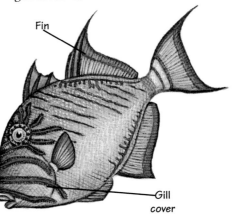

Gill cover

gills 11, 15. *Breathing parts of fishes and many other water animals.*
**Great Barrier Reef** 5, 6, 31, 34, 36, 39
grouper 25

**H**
hard coral. *See coral animal.*
Hawaii 38, 39

**I**
Indian Ocean 39
ink 6
invertebrate 41, 42

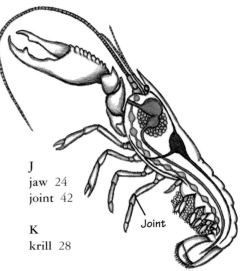

**J**
jaw 24
joint 42

Joint

**K**
krill 28

**L**
lagoon 6. *Shallow water separated from the open ocean usually by a coral reef.*
leg 42
limestone 8, 11, 13, 22, 28
lionfish 31

**M**
magnifying glass 13, 31
mammal 40
mantle 15. *The part of clams, snails, and many other water animals that gives off the chemicals that harden into the shell.*
marine park 36
microscope 43. *An instrument that makes tiny objects look many times larger so that people can see them.*

# Index

mineral 10. *A kind of chemical that, in living things, helps a cell work the way it should. Calcium and iron are two examples of minerals.*

mining 36

monera (muh-NEER-uh) 5, 43. *Creatures made up of one cell that doesn't have a nucleus (control center).*

moon 32

mouth 9, 14, 15, 16, 17, 21, 23, 28, 31, 32

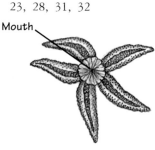

Mouth

mucus 20, 21, 25, 29

muscle 23, 30, 31

Muscle

Clam shell

mushroom coral 33

**N**

nerve 9

notebook 9, 10, 17, 25, 31, 32

**O**

ocean 5, 6, 8, 24, 38, 40

octopus 15, 18

overfishing 36

oxygen (AHK-suh-jin) 9, 11, 12, 15. *A gas in the air and water that living things breathe.*

**P**

Pacific Ocean 39

parrotfish 14, 25

pearlfish 12

pest 17

plankton 24, 28, 31, 34, 43. *Tiny animals, eggs, algae, and other living things that float in the water.*

plant 5, 12

poison 15, 16, 19, 21, 29, 30, 31

pollution 36

polyp (PAHL-ip) 8, 29, 32, 34

predator (PRED-uh-tur) 10, 19, 20, 21, 24, 25, 31, 32, 34. *Animal that kills other animals for food.*

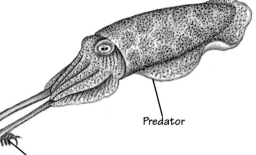

Predator

Prey

prey 15, 19, 31, 34. *Animal hunted or caught for food by a predator.*

protist 5, 43. *A living thing usually of one cell that has a nucleus (control center).*

pufferfish 16

**R**

Red Sea 39

reptile 40

rock 5, 8, 9, 10, 12, 19, 21, 22, 29, 33, 36

**S**

sailing 5, 36

salt 40

sand 14

school 22, 24

scorpionfish 16

sea. *See* ocean.

sea anemone 9, 20, 21, 25, 26

sea cucumber 12, 30, 42

Sea of Cortez 38

sea slug 26

sea squirt 22, 23

sea star 38, 40

sea turtle 38, 40

sea urchin 12, 22, 42

seawater. *See* water.

seaweed 12, 13, 22

shark 6, 24, 34

shell 14, 19, 21, 26, 31, 33, 34, 42

shrimp 10, 12, 13, 16, 20

skeleton 10, 11, 21

# Index

Shark teeth

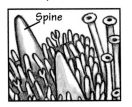
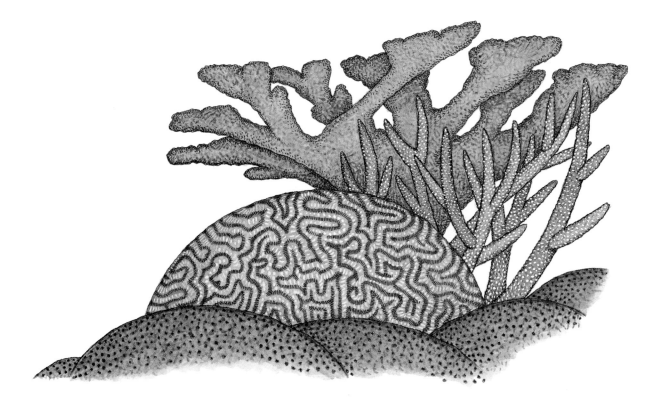

# Find Out More

*Look for the following in a library or bookstore:*

**A Field Guide to Coral Reefs** by E. Kaplan, Houghton Mifflin Co., Boston, MA

**Reef Creature Identification** by P. Humann, Vaughan Press, Orlando, FL

**Fishwatcher's Guide to West Atlantic Coral Reefs** by C. Chaplin,
Harrowood Books, Newtown Square, PA

**Indo-Pacific Coral Reef Field Guide** by G. Allen and R. Steene,
Tropical Reef Research, Singapore

**Australia: Seas Under Capricorn**, Time-Life Video, 1990

**Splendors of the Sea: The Caribbean's Secret World** (video), The Discovey
Channel, 1992

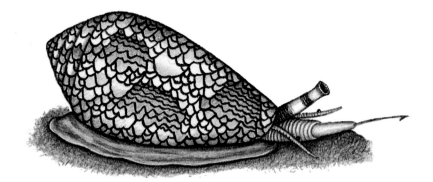